How to Install Electric Bells, Annunciators and Alarms

HOW TO INSTALL
Electric Bells, Annunciators, and Alarms.

INCLUDING

Batteries, Wires and Wiring, Circuits, Pushes, Bells,
Burglar Alarms, High and Low Water Alarms,
Fire Alarms, Thermostats, Annunciators,
and the Location and Remedying
of Troubles.

BY

NORMAN H. SCHNEIDER,

Author of "The Study of Electricity for Beginners," "Care and
Handling of Electric Plants," etc., etc.

SECOND EDITION, ENLARGED

NEW YORK
SPON & CHAMBERLAIN, 123 LIBERTY STREET

LONDON
E. & F. N. SPON, Limited, 57 HAYMARKET, S.W.
1913

The Camelot Press, 16-18 Oak St.. New York

PREFACE

Among all the applications of electricity to domestic or commercial uses, few are as widespread as the electric bell. Practically every building used for a dwelling, storage or manufacture requires an electric bell, annunciator or alarm system.

This book was written to explain in practical language how an electric bell system operates and how it is installed; its success shown by its large sale has resulted in this new edition which brings the subject up to date.

Many new diagrams of annunciator and burglar alarm systems have been added, together with descriptions and illustrations of wiring elevators for electric bells, wiring for door openers, the use of transformers for furnishing suitable ringing current from electric light circuits; and high voltage bells intended to be used on other than the customary low voltage battery circuits.

The author expresses his acknowledgment to the Western Electric Company for diagrams of door opener circuits in connection with their interphone systems, to Edwards and Company of New York for diagrams of fire alarms, burglar alarms and annunciators, and to the Westinghouse Company for illustrations of bell-ringing transformers.

CONTENTS

Chapter IV

Chapter V

Chapter VI

LIST OF ILLUSTRATIONS

LIST OF ILLUSTRATIONS

INTRODUCTION

An electric bell depends for its action on the fact that a piece of iron wound with insulated wire becomes a magnet and will attract another piece of iron just so long as an electric current is allowed to travel through the wire.

The instant the current ceases, the magnetism also ceases, and the attracted piece of iron (termed the armature) is no longer held in contact.

The general construction of an electric bell is shown in Fig. 1. $M M$ are coils of insulated wire wound on soft iron cores. A is a soft iron armature mounted on a flat spring so that it is normally kept a slight distance away from the soft iron cores. S is a brass screw with a platinum tip touching a platinum disc on a spring attached to the armature.

When the push button P is pressed down, its two brass springs touch each other, the current from the battery cell B then flows through the wire W, through the push P, through the coils $M M$, along A to the platinum disc, out

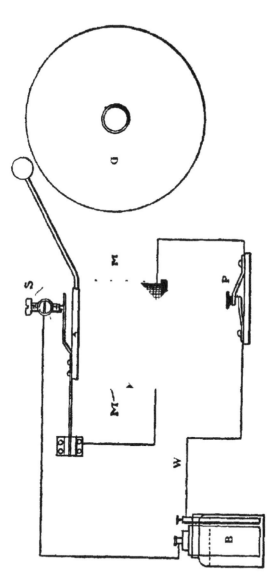

Fig. 1

at *S*, which touches this disc, and back to the battery.

The instant this is done the current causes the iron cores to become magnets, they attract *A*, which then breaks contact at *S*. The spring mounting of *A* causes it to jump back to its first position, *S* then touches the platinum disc again, the current flows as before, and the armature is again attracted only to break contact with *S* and fly back.

This continual making and breaking of the circuit keeps up as long as the push is pressed, a ball mounted on *A* by means of a rod strikes against the gong *G* causing a continuous ringing of the bell. The wires leading between the bell, battery cell and push must all be insulated, that is, covered with cotton, rubber, etc., which prevents the leakage of current should two wires cross each other. Copper wire is mostly used for circuits indoors, the details of the kind and size of wire will be given later on.

The main parts of an electric bell circuit are then—the battery to supply the electric current; the circuit, or wires, to carry this current; a push, or circuit breaker, to control the current flow; and a bell to utilize the current.

CHAPTER I

The Battery

The Battery Cell. The battery cell most used in electric bell work is the Leclanche, or some modification of it.

The Leclanche battery cell is shown in Fig. 2,

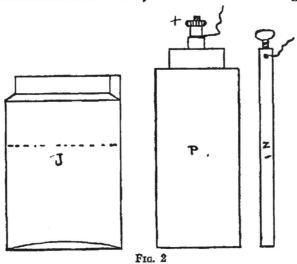

Fig. 2

where J is a glass jar, Z a rod of zinc, and P a jar of porous earthenware containing a carbon rod surrounded by powdered carbon and peroxide of manganese.

In setting up this cell about four ounces of sal ammoniac (chloride of ammonia) are put into the jar and enough water added to come about half way up the jar.

The porous jar P and the zinc Z are then inserted, and the cell is ready for use in a few minutes after the liquid has soaked through the earthenware into the carbon-manganese mixture. Water is often poured into the porous jar through holes in its top to hasten this wetting.

Wires are clamped by nuts or set-screws to the negative terminal on the zinc or the positive terminal on the carbon, it generally not being of consequence which terminal is attached to either wire of the circuit.

A battery cell could be constructed without the manganese, using simply a plate of carbon and a rod of zinc, but hydrogen gas would be generated on the carbon plate when the cell was working and would stop the current flowing.

This is called polarization, and peroxide of manganese is a de-polarizer, because it combines with this hydrogen gas almost as fast as it is generated, and prevents, to a great extent, the polarization.

But it does not stop it entirely, as will be seen if the Leclanche cell is kept working above its capacity. Then the hydrogen is generated too fast for the manganese to destroy it, and the cell ceases to work. In this case a rest will often restore the cell to its former power.

Cells which have been almost unable to make a bell give even a single tap have been found good again when allowed to remain at rest over night.

In setting up a battery cell no liquid should be splashed on the brass terminals or corrosion will take place. Every metal surface where connection is made to allow electric current to pass *must* be clean and bright, and all screws, or nuts, holding wires must be screwed up tight so that the wires are firmly clamped.

Loose or dirty connections are the cause of probably eight out of every ten troubles affecting bells and batteries.

When the fluid in a Leclanche cell becomes milky, more sal ammoniac must be added. Or, better still, throw out the old solution, wash the porous jar thoroughly in clean water, scrape the zinc bright, and half fill the cell with fresh solution.

The zinc wearing away rapidly or becoming covered with crystals, and a strong smell of ammonia, show generally that the cell is being worked too hard, or that the current is leaking where it should not.

A zinc rod in a cell working the average door

bell should last for six months, the porous jar for
a year.

The Dry Cell. The Leclanche cell being a
cell with much free liquid is liable to dry up if
not watched. The dry cell (Fig. 3) is a modern

FIG. 3

form of the Leclanche where the liquid is held by
an absorbent material, such as blotting paper, or
plaster.

A typical dry cell* is shown in the figure. An

---

*For full description of this class of battery see No. 3
Book on "Dry Batteries."

outside case of zinc is lined with blotting paper dampened with chloride of zinc and sal ammoniac. A carbon rod is then inserted in the centre and packed around with carbon dust and peroxide of manganese. The latter mixture is also somewhat dampened.

Molten wax, or a suitable composition, is then poured on top of the contents of the cell to seal it up and prevent the evaporation of the fluid. A

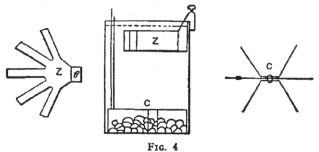

FIG. 4

terminal on the carbon rod and another on the zinc case complete the cell.

The voltage of both the Leclanche and the dry cell is about 1.45, when it goes below this it indicates that the cell is worked out.

The two cells described are known as open-circuit cells and are only intended for intermittent working.

When a current is needed for a long period at a time a closed circuit cell should be used, such as the gravity Daniell cell.

The Gravity Daniell Cell. The gravity cell, Fig 4, has a zinc block Z suspended from the side of the jar and a number of copper leaves C standing on edge at the bottom. A quantity of bluestone (sulphate of copper) is poured over the copper leaves and the jar filled with water.

During the working of this cell, copper is deposited on the copper plate, and sulphate of zinc formed at the zinc. To hasten the action a small quantity of zinc sulphate can be added to the solution when setting up the cell

The name of this cell comes from the fact that the copper solution being heavier remains at the bottom of the jar If the cell is not worked enough, all the solution will become blue and the zinc will blacken. If very dirty from this cause, remove the zinc, scrape and wash it thoroughly. Throw out all the solution, add new sulphate and water and replacing the zinc, then put the cell on short circuit by connecting the copper and zinc together for a few hours.

E. M. F. The e. m. f. of a gravity cell is within a fraction of one volt, its current nearly one-half ampere.

Warmth makes it give a greater current; on no account let a gravity cell freeze.

Resistance of a Cell. The fluids in a cell do not conduct electricity as well as copper does; they offer more resistance and thus reduce the current output.

The internal resistance of a cell may be lowered by using large zinc plates curled around the porous pot.

The Samson cell has a large zinc plate bent in the form of a cylinder, the carbon-manganese combination standing in the centre of it.

The dry cell also has a large zinc, the internal resistance being thus much lowered, the current output is increased. This is by reason of Ohm's law, which teaches that to increase the current flow, either the voltage of the battery must be increased, or the resistance decreased.

But increased current means lessened life; there is only just so much energy in a cell mainly dependent on the quantity of chemicals.

Grouping of Cells. Cells may be grouped in a battery to get increased voltage, or increased amperage. When connected for the former, they are in series, the carbon of one is connected to the zinc of the next, and so on.

If all the carbons are connected together and all the zincs, they are in multiple, and will give the same voltage as of one cell but the combined amperage of all.

In ordinary bell work the series is the general connection, the higher the resistance of the circuit, or the longer the wires, the more voltage is required.

CHAPTER ·II

Bells and Pushes

Electric Bells. The two main types of house bells are the iron box and the skeleton.

The iron box has a cast-iron frame, or base, and a cast- or stamped-iron cover over the mechanism.

The skeleton bell has an iron frame but no cover, and is generally better finished and more expensive than the iron box bells.

For fire alarm purposes, mechanical bells or gongs are made, in which a clockwork mechanism causes the hammer to strike the gong upon being released by electromagnetism.

Marine or waterproof bells have an iron cover fitting tight over a rubber gasket; they are for marine, or mining, work.

Polarized, or magneto, bells are used in telephone work, and are rarely operated by a battery, but have a miniature dynamo generator operated by hand, or power, to supply the actuating current.

Most bells are classed for size by the diameter of the gong, a four-inch bell being one with a gong four inches in diameter; a six-inch bell one with a six-inch gong, and so on.

According to the use for which they are intended, bells may be vibrating, as before described, single-stroke, shunt or short-circuiting, differential, continuous-ringing, or adapted for circuits of high voltage.

The Single-stroke Bell. The bell before described, and again shown in Fig. 5, is a vibrating, or trembling, bell. It is often desired to have the hammer give only one stroke for each pressure of the push, as in signaling with a code of taps; in this case a single-stroke bell is used. The circuit

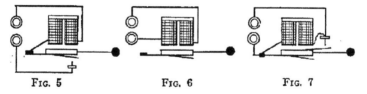

FIG. 5   FIG. 6   FIG. 7

from the binding posts is then directly through the magnet coils without any break at the contact screw, as in Fig. 6.

In adjusting such a bell to give a clear sound, press the armature up against the iron magnet cores and then bend back the hammer until it just clears the gong. The spring of the hammer wire will carry the hammer sufficiently forward to hit the gong. The tone will be clearer than if the hammer dampered the gong by pressing against it when the armature was nearest the core.

By bringing out a third connection, a vibrating bell may be made both single stroke and vibrating.

The Shunt Bell. There is a form of bell, Fig. 7, known as the shunt, or short circuit bell, which is often used when two or more are to be connected in series, as will be seen in the description of circuits. In this bell the circuit through the magnets is not broken at the contact screw, but the forward movement of the armature short circuits the coils.

As the short, or shunt, circuit is very much lower in resistance than the wire on the magnet coils, the main current flows around the latter and they do not become energized. The sparking at the shunting contact screw is much less than it would be at the ordinary breaking contact screw, and the platinum points last longer.

The Differential Bell. Sparking at the breaking contacts of an electric bell is detrimental to the platinum points, and many remedies have been devised to overcome it.

Sparking is due to the self-induction of one turn of the wire coil acting on its neighbor, and this property is utilized in the gas engine, or gas-lighting spark coil, where a fat spark is needed to ignite gas.

The differential bell has two windings in opposite directions. The action of one would be to produce an N-pole at one end and an S-pole at the other. But the second coil produces poles just the opposite, as the polarity of a magnet depends on the direction in which the current flows around it.

Where the current flows around the first winding the armature is attracted and its spring contact meets the contact screw and allows the current to divide, part flowing through the first coil, the other flowing in the reverse direction in the opposite way. One coil would tend to produce an N-pole where the other coil produced an S-pole, and these opposite poles would so neutralize each other that there would be no magnetism.

The armature would therefore be pulled back by its spring when both coils were thrown into circuit. In so doing it would cut out one coil and the same series of operations would recommence.

As a spark is normally produced where magnetism is *lost* by a break of circuit,* no spark appears, as magnetism is *produced* by a break of circuit in this case.

---

*For a full explanation of self-induction see No. 1 of this series.

Continuous-ring Bell. In some classes of bell work, such as burglar alarms, it is desired that the bell when once started shall continue to ring until stopped by the person called. In this case a continuous-ringing bell is needed, such as in Fig. 8.

When the push *P* is pressed, the current flows

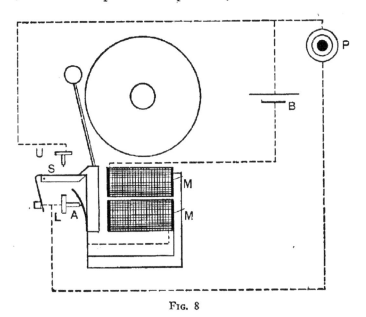

FIG. 8

in the usual way through contact screw *L,* armature spring *A,* magnet coils *M M,* battery *B,* back to *P,* and the bell rings. But on the first forward movement of the armature it releases the spring contact *S,* which flies forward and makes contact at *U.* The circuit is now from *B,* through *M M,*

to *A*, thence through *L* and *S*, to *U* and back
to *B*.

The bell will continue to ring until the spring
contact *S* is moved back and caught by the projection on the armature *A*.

A continuous-ring attachment is also made and
sold in most electrical supply stores, which is complete in itself and can be applied to any bell.

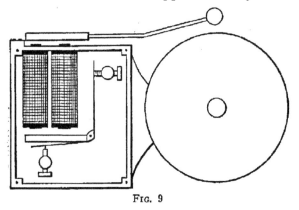

FIG. 9

Waterproof Bells.   In Fig. 9 is an example of
a waterproof bell where the mechanism is almost
all entirely encased in a waterproof brass case.

The circuit is made and broken inside the case,
but the magnet cores project through it and act
on a second armature placed outside. This second armature carries the hammer which strikes
the gong and is governed in speed by the contact-
breaking armature inside.

Forms of Bell Gongs. In order to provide a variety of sounds, bells are provided with gongs of various shapes.

Fig 10 shows the ordinary form of gong.

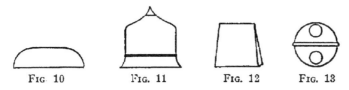

Fig 10   Fig. 11   Fig. 12   Fig. 13

Fig. 11, a tea gong; Fig. 12, a cow gong; and Fig. 13, a sleigh bell.

A coil of steel wire is also used, as in Fig. 14, which on being struck by the hammer gives a pleasant but not loud tone.

Fig 14

The Buzzer. The buzzer is the mechanism of a vibrating bell less the hammer and gong. As the armature vibrates it makes a buzzing noise which does not carry as far as the sound from a struck gong. It is used chiefly for a desk call and in telephone exchange work, or any place where general attention is not desired to the signal.

Operating Bells at a Distance. When it is desired to ring a bell situated at a considerable distance from the push, the resistance of the line becomes objectionable.

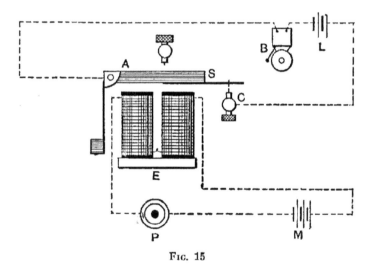

Fig. 15

On lines of 500 feet, No. 18 copper wire and upwards, the battery necessary would be very large, two small batteries and a relay would prove more satisfactory.

In Fig. 15 the circuit of a simple form of relay is given. An adjustable contact screw C is placed where an extension S of the armature A can strike it. This extension is provided with a platinum contact. The connections are as in the figure.

When the push *P* is depressed, the current from the main battery *M* energizes the electromagnet *E*, and the armature *A* being attracted, contacts *S* and *C* meet. These contacts close the second circuit containing the bell 3 and the local battery *L*.

The relay resembles a second push near the bell, but controlled by current from a distance instead of being depressed by hand. Its advantage consists in it needing but a very weak current to move the armature *A*, which is held back by a light spring, or by gravity.

The relay may then be set near the bell and the wires from the push may be of a very great length. Battery *L*, which actually rings the bell, will thus only have to work through a few feet of wire.

Reducing Resistance of a Bell. Sometimes it is desired to reduce the resistance the bell coils offer to the current, the bell then working over a very short line with few cells of battery. Or the bell coils may have been wound with fine wire for large battery voltage and a long line.

The bell coils may be put in multiple, the current then dividing and one-half going through each spool.

Untwist the joint between the spools near the yoke or iron bar to which the spools are attached. Join one of these ends to the wire at the armature end of the *other* spool and the second untwisted end to the armature end wire of its neighboring spool. Use short pieces of insulated wire for these extra connections.

The current now instead of having to go through one spool and then the other, can branch through both at once.

The resistance to the current of one spool is half the resistance of two, the current through one spool will therefore be twice that through the two spools as at first connected. And as there are two paths for it, each one-half the first resistance, the total will be only one-fourth the resistance of the ordinary series arrangement.

The same size battery will therefore send four times the current through the spools in multiple than when they are in series.

It is to be noted that the wire on one spool is wound in the reverse direction to that on the other. The reason will be apparent if the two spools and yoke are considered as merely one spool bent in a U or horseshoe form.

If both spools were wound in the same direction they would be in opposite directions when the U were straightened out, and would cause like_

poles at the same ends   These poles would neutralize one another, so that there would be no
magnetic attraction.

This can be readily proved by joining together
the two yoke ends and the two armature ends of
the spool wires.   Then pass the current through
these two joined connections.

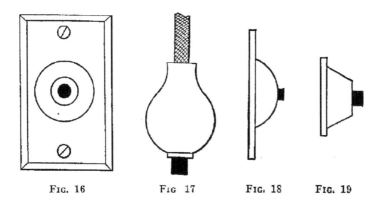

FIG. 16   FIG 17   FIG. 18   FIG. 19

The Push Button.   Push buttons, or pushes,
are made in a variety of forms, with metal, wood,
hard rubber, or porcelain bases.

Fig. 16 has a metal base, and is suitable for
a front door.

Fig. 17 is a wooden pear push, and is attached at
the end of a cord which has the two conductors
braided in it, each, however, having its own insulation.

Fig. 18 is a plate push for an outside door.

Fig. 19 is either of metal, wood, or porcelain, and is the shape most commonly used.

A three-point push has three contact springs. One is movable by means of the button, one is below the movable spring, and the third is above it.

When the push button is not being depressed,

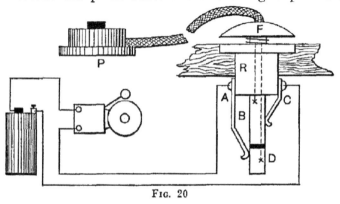

FIG. 20

the movable spring makes contact with the upper spring. But when the button is depressed, these two springs part, and the movable spring makes contact with the lower one.

This style of push is used for special bell and annunciator work, as will be described later.

The form of combination floor and table push in Fig. 20 is the most solidly constructed device of its kind. The lower part is set in a hole bored in the flooring, the metal flange keeping it in place and preventing its slipping through.

The floor push attachment works as follows: The central metal rod is divided into two parts *B D,* by an insulating piece of hard rubber. When depressed against the action of the spiral spring by the foot, the upper part *B* connects together the contact springs *A C,* closing the circuit of bell and battery. These contact springs are insulated from each other by a hard rubber block *R.*

From the table push a cord containing two insulated wires leads to the two parts of the rod at *B* and *D.* When the push centre is pressed down, the push springs come together and practically short circuit *B* and *D,* which completes the circuit of bell and battery. At any time the centre rod may be removed, leaving a surface almost flush with the carpet, or floor, over which furniture may be moved without injury to the mechanism of the push.

For a floor push alone a shorter form of the centre rod is also sometimes furnished which is not divided by insulation. The spiral spring keeps it clear of the lower contact *A* but enables it to always make connection with the upper contact *B.* Pressing this rod down will also short circuit the bell and battery so that the signal is given.

A door pull attachment, like Fig. 21, is made so that the ordinary form of lever pull bell may

be changed into an electric bell. Being screwed
up near the door pull, a wire is run from the latter and fastened to lever *L*. When the pull is
drawn out the lever *L* turns on a pivot and a
projection presses the insulated spring *S* against
the metal base *B*. The circuit of the bell and
battery being thus closed, the bell rings.

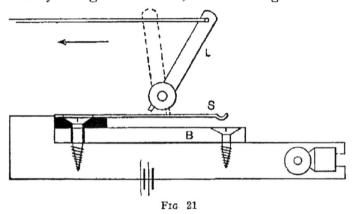

Fig. 21

Indicating Push Button. A push button is
made which contains in the base a small electromagnet in series with the line. An armature on
a spring is fixed near the magnet poles. When
the push is depressed, the current travels through
this electromagnet, and as the circuit is made
and broken at the distant bell, it is also interrupted
in the electromagnet. The armature vibrates in
unison with the bell and thus gives an audible
indication that the bell is ringing.

CHAPTER III

Wiring, Circuits and Troubles

The Wire. The size of the copper wire used in bell work is No. 16, or No. 18, B and S gauge, and sometimes smaller, such as No. 20 to 22. But smaller wire than No. 18 has too much resistance, and would necessitate a larger battery power, even if its mechanical strength were not too low. The insulating coverings are cotton saturated with paraffin wax or compounds.

The covered wires are variously known as annunciator, office, or weatherproof wire, these terms being mostly for distinction of the coverings and not for the use to which the wire would be put.

Annunciator wire has two layers of cotton merely wrapped around the copper and then saturated with paraffin.

Office wire has the two cotton layers braided, the inside one being filled with a moisture-repelling compound.

Both office and annunciator wires have their outside coverings filled with paraffin and highly polished.

From the ease with which annunciator wire is

stripped of its cotton covering, the braided office wire is to be preferred. These coverings are made in a variety of colors.

Weatherproof covered wire is mostly used for electric light work, but the sizes given above are good for bell work, although their larger outside diameter makes them harder to conceal.

The approximate number of feet to the pound of office and annunciator wire is given in the table.

Office Wire		Annunciator Wire.	
No	Feet per lb	No.	Feet per lb
12	35	18	180
14	55	20	225
16	95		
18	135		

Joints.   Upon the care with which a joint is made much depends, a loose or poorly made joint will offer much resistance to the current.

The correct way to start a joint in annunciator, or office, wire is shown in Fig. 22. About three inches of each wire to be joined is bared of its insulation and scraped bright. The ends are then

bent at right angles to each other, hooked together
and one end firmly twisted around the other, as
shown in Fig. 23 Any projecting pieces are cut
off, and the joints should then be *soldered* to prevent corrosion.

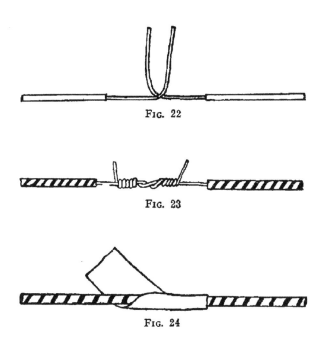

FIG. 22

FIG. 23

FIG. 24

Adhesive tape ("friction tape") is wrapped
around the joint, Fig. 24, and pressed firmly together so that there is no chance of its unravelling.
The tape wrapping should extend across the joint
and on to about a half inch of the insulation
around each wire.

Running the Wires. To detail all the operations of installing a complex system of bell, alarm and annunciator wires would be impossible from the reasons that conditions vary and space is limited. General directions will then only be given to enable the inexperienced to run such wires as may be needed in ordinary domestic work and to guard against the most common causes of failure.

Wires may be run in tin tubes to prevent the depredations of rats and mice, or they may be run with simply their own covering for protection; it is presumed the latter is undertaken.

In a case where the building is of frame and in course of erection the task is much simplified.

Having first decided upon the plan, number of bells, pushes, etc. and their location, proceed to run the wires first in order that the pushes, bells, etc. may not be injured.

But where the house is already occupied, as in the majority of cases likely to be met with by the reader, the bell and battery may be set first.

Take the case of an ordinary door bell with the push at the front door, the bell in the kitchen and the battery in the cellar. If possible get the wire on two spools; it will simplify matters if both wires are of different colors. Starting at the push, have a foot of each wire for connection and slack, and fasten each wire lightly to the woodwork with staples, or double-pointed tacks, never putting two wires under one staple nor driving in a staple so it cuts the insulation. Some cases will require a staple about every foot, on straight runs sometimes every three feet.

In many cases the wires can be partly concealed in the angle between a moulding and the wall, or even in a groove of the moulding itself. When running along a skirting, the wires may often be pushed out of sight between it and the floor. Do not attempt to draw the wires too tight or the changes of the weather may break the wires when the woodwork shrinks or swells.

The wires will be, one from the push to the bell, one from the push to the battery, and one from the bell to the battery. So it is probable that the second wire can be run right through a small hole bored in the flooring under the push, but inside the front door. In this case it will be perhaps easier if the spool be left in the cellar and the end of the wire be pushed up from below and stapled to the woodwork near the push, leaving the cellar work to the last. Only one wire will be run then direct to the bell upstairs and it can be better concealed than two.

If necessary it may be drawn under a carpet and not stapled, or it can often be forced into the crack between two boards. But if not, run it along the skirting, following the walls until it reaches below the bell. It is often better to go entirely around a room than to cross below a door.

If a door must be crossed the wire may either run up one side of the frame and down the other or laid beneath the carpet on the sill. The former is preferable, but takes more wire.

In many houses the bell wire as well as the battery wire may be run across the cellar beams (Fig. 25), in which case bore a second hole for it near the push; do not draw it through the same hole as the push to battery wire. And, of course, here work upwards with the spool in the cellar.

Having reached the bell location, run the third wire down into the cellar to the battery. Now connect up the push, baring an inch or so of each wire, push them through the holes provided in the push base, screw down the push base and clamp the wires under the washers through which the connection screws run. Do this neatly, be sure the ends of the wires do not stick out, cut off what is left free of the bared ends. Then connect the battery to the wire from the push and the wire from the bell. The last thing is to scrape and fasten the bell wires to the bell binding posts. Do this so that they cannot come loose and that they make good contact.

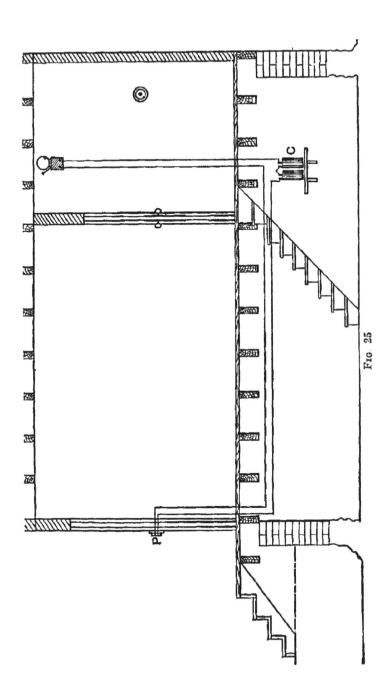

Fig. 25

The bell should now ring properly when the push is pressed.

To sum up, one wire leads from one spring of the push to the bell, one wire from the other spring of the push to the battery, and another wire from the remaining binding post on the bell to the remaining binding post on the battery. It is immaterial whether the zinc terminal or the carbon terminal go to the bell or push.

Combinations of Bells and Pushes. One of the wires in a bell circuit may be replaced by the ground (Fig. 26). Connection may be made to a gas or water pipe or to a metal plate buried deep in damp earth. Any wire fastened to such a

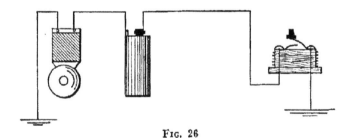

FIG. 26

plate must be thoroughly soldered to it or a voltaic action will be set up, which will eat it away at the point of contact.

When one bell is to be rung from two or more points the pushes are to be connected in

multiple (Fig. 27) as if they were in series; all
would have to be closed to complete the circuit.

If two bells are to be operated from one push

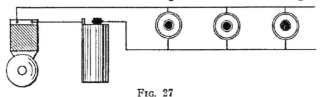

FIG. 27

they may be in series (Fig. 28), but in this case
one of them must be arranged for single stroke.

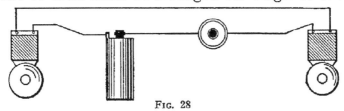

FIG. 28

If both were vibrating bells the armature of one
would not vibrate in unison with the other arma-

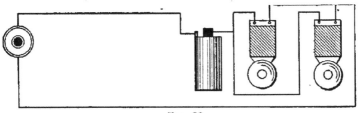

FIG. 29

ture and the result would be irregular contact
breaking and intermittent ringing.

A preferable connection for two or more bells

and one push is Fig. 29, where the bells are in
multiple. This requires more current than the
series method.

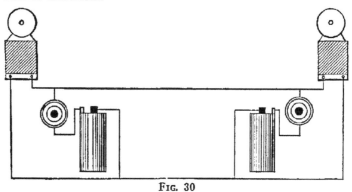

Fig. 30

To ring two bells from either one of two points,
the arrangement in Fig 30 will answer. It requires only two wires or one wire and ground
return, but two batteries. As both bells are in

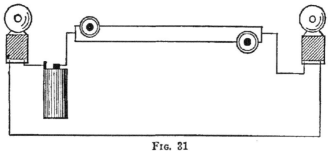

Fig. 31

multiple both will ring, the one nearest the push
being depressed ringing the loudest. This is a dis-
advantage. If the series arrangement in Fig. 31

be selected, one bell must be arranged for single stroke.   Both bells will ring with equal power.

In Fig. 32 only the distant bell rings, the cir-

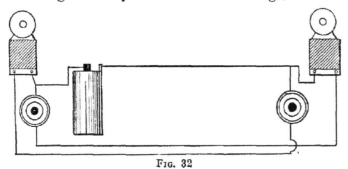

Fig. 32

cuit having only one battery but three wires, or two wires and ground return.

A plan where two batteries are needed but only two wires, or one wire and ground is in Fig. 33. Double contact or three-point pushes are necessary

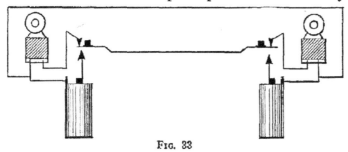

Fig. 33

here, making one contact when depressed and a second one when not being touched.

In this figure only the distant bell rings.

Faults in Bells. On examining many electric bells it will be noted that only one binding post is insulated from the frame when the latter is

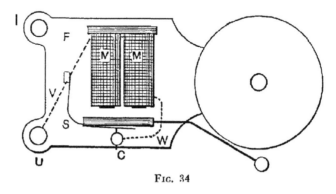

FIG. 34

of iron (Fig. 34). As the armature spring S is in electrical connection with the frame F by reason of its metal screws and support, the circuit may run from the insulated post U to the magnet coils, thence through the insulated contact screw C through the armature spring (when it is making contact) and through the frame to the uninsulated post I.

This saves labor, wire and complication, but if the insulation of the post U, the wires W V, or the contact screw C be injured, the current may take a short path back to the frame.

If C were thus grounded, the bell would act as a single-stroke bell.

If U were grounded, the bell would not ring

at all, as that would be a short circuit on the battery between I and U and the latter would also result if the bare wire were touching the frame at V.

If the bare wire touched the frame beyond M M, that is, along W, it would be a single-stroke bell, as if C were grounded.

As any one of these faults is likely to occur, they should be looked for when the bell acts imperfectly, or not at all.

A very common fault in a bell is when its armature sticks to the cores and thus does not make contact with the contact screw. This may be from a weak spring or because of the loss of the pieces of brass inserted in the ends of the cores to keep the armature away from actual contact. A piece of a postage stamp stuck over the core end will often help out in the latter case.

A high screeching noise from the armature vibrating too rapidly but with too little play, may be from excessive battery power or the contact screw being too far forward. The former will generally be detected by the violent sparking as well as the rapid vibration.

In very cheap bells the platinum contacts may be replaced by German silver or some other metal.

Platinum is necessary because the sparking would soon corrode other metals, but it is very expensive. To test for platinum put a tiny drop of nitric acid on the suspected metal. If bubbles or smoke appear it is not platinum. After applying this test in any case however, carefully wash off and remove all traces of the acid, as it will corrode the metal into which the platinum is riveted.

Dirty contacts will decrease the current in the bell coils and it will not work well, if at all.

Loose contact screws and wires also give trouble. The adjusting of the contact screw is of the utmost importance, and should never be attempted unless it is clearly necessary.

Faults in Line. In looking for a fault in a bell circuit make sure the battery is working; if only one or two cells, put the ends of two wires attached to the terminals on the tongue: a metallic taste will indicate current.

Then see that the circuit wires are firmly clamped in the terminals and no dirt or corrosion on the connections.

Next examine the push button and see that the wire connections at the springs are perfect.

If there is no movement of the bell at all when the push is pressed in, take a pocket knife or screw driver, and touch the blade across the push springs. If there is current flowing sparks will be seen when the blade breaks contact between the springs. If there are no sparks, detach the wires from the bell and twist the bare ends together. Then try again for sparks—they may now be very minute. The tongue test is good here.

If current is detected, examine the bell for the defects first mentioned.

But if no current is found at the push now the wires are broken somewhere.

First short circuit the push springs by inserting a knife blade or piece of wire so as to touch both of them. Then touch the two wires at the bell, one to each side wire coming from the magnet coils. If current is up to the bell and the coils are all right, a single stroke should result.

Replace the wires in the binding posts, clean the platinum on both contact screw and armature spring and try the adjustment. Troubles in the bell will be mostly similar to those before mentioned.

If no current has been obtained at either bell or push, and the battery is in good working order, the line must be tested for a cross or break.

If the wires are touching each other (Fig. 35) at some bare spot S between the bell and the battery, it will be shown by the metallic taste upon detaching one wire from the battery and laying it on the tongue T, together with another wire W from the disconnected terminal of the battery. The

current will travel from the battery to the cross
at *S*, then back along the second circuit wire to
the tongue and through the short wire to the
battery.

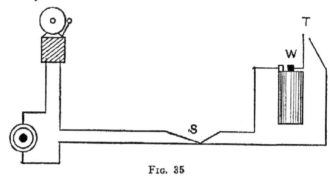

Fig. 35

If no current is obtained in this way it is probable that the wire is broken.

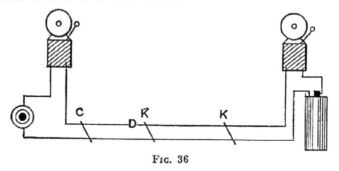

Fig. 36

The easiest way to find this is to take a bell
to the battery and connect it between the circuit
wires and the battery (Fig. 36).

Then with a sharp knife carefully cut away a

little piece of the insulation from each wire beyond the bell and battery and short circuit the bared spots with the knife blade *K*. Keep working towards the push. The bell will ring each time

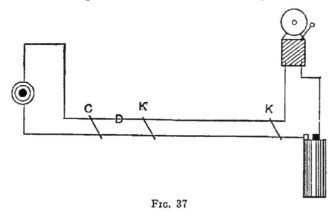

Fɪɢ. 37

at *K K* until the break *D* is passed, at *C* it will not. It becomes an easy matter then to locate it.

If the bell and push are far apart, as in Fig. 37, a break between the push and the bell may be found as shown. With the knife blade *K* at different points the bell will ring, but after passing the break *D* it will not ring.

Such simple tests as are here given can be carried out by any one, but far better results will be obtained if the reason for each is first learned.

This can be readily done by a careful study of the diagrams and text.

CHAPTER IV

Alarms

Fire Alarms. Thermostats, heat alarms and fire alarms are all practically the same, the term thermostat being applied principally to the apparatus which closes the electrical circuit.

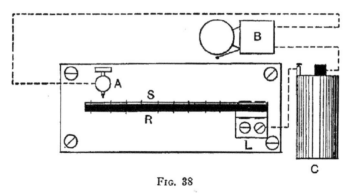

FIG. 38

Thermostats act on the principle that heat causes expansion whether of substances, liquids, or gases.

The degree in which different substances expand varies for the same increase in temperature. This fact is used in a common form of thermostat shown in Fig. 38. A strip of wood or hard rubber *R* has a strip of thin sheet metal *S* riveted to it. This compound strip is held at one end by

a lug *L* screwed fast to a baseboard. Upon an
increase of temperature the hard rubber expands
more than the metal strip and the compound strip
bends towards the adjustable contact screw *A*.
Upon touching the latter, the circuit through the
bell *B*, battery *C* and the metal strip *S* is com-
pleted, and the bell rings. A contact screw can
be arranged at the other side of *S R*, which will

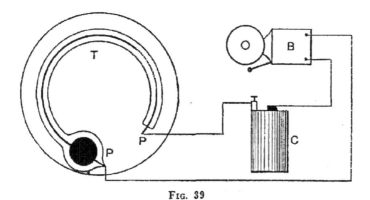

Fig. 39

give warning of a decrease in temperature, as the
rubber contracts more than the metal strip.

In some thermostats of this character two metals
having different coefficients of expansion, such as
steel and brass, are used instead of metal and hard
rubber.

Thermostats of this nature are much used in
incubators, and they can readily be combined with
electric apparatus to open or close hot-air valves,

dampers, etc., and thus regulate the supply of hot air, hot water, or gas.

A thermostat much used in fire alarm work has a thin metal chamber which is air tight. An increase of temperature causes the air to expand, which swells out the walls of the chamber and closes an electric circuit.

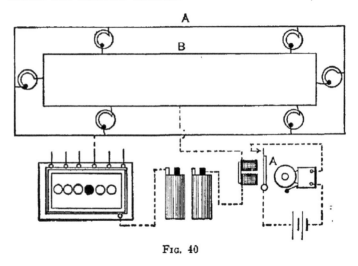

Fig. 40

The mercurial thermostat shown in Fig. 39 has a glass tube T and bulb containing mercury. Into each end is sealed a platinum wire PP. Upon the temperature rising to a predetermined degree, the expanded mercury completes the circuit between PP and the battery C and bell B are put in operation.

Fig. 40 is the open circuit system most used by

the fire alarm companies, only one circuit of six thermostats being illustrated.

It will be seen that if any thermostat closes the circuit between the outer and inner wires of the ring *A B,* current will flow through the corresponding drop of the annunciator and will attract the armature *A* of the relay. This will cause the bell to ring. As the relay is connected to the annunciator as before shown for the annunciator bell, it offers a common path for any drop to the battery. Thus the bell will ring for any circuit, but the individual drop only will fall. In a simpler circuit the relay may be dispensed with and a vibrating bell only used.

Thermostats may be operated on open or closed circuits, that is, they may give the alarm by closing a circuit and ringing a bell, or by opening one and releasing a contact spring as in the burglar alarm system to be described later.

Water Level Alarms. Where it is desired to signal the rising or falling of water in a tank above or below a given point, a water level indicator as in Fig. 41 may be used.

A hollow ball *H.* is mounted on the end of a rod which slides vertically in guides, not shown. Adjustable stops *S S* press against a spring arm *R,* pressing it up or down, according as the water level is rising or falling. If rising, R makes contact with the adjustable screw A, if falling, with D, in both cases completing the electrical circuit of the battery C and bell B.

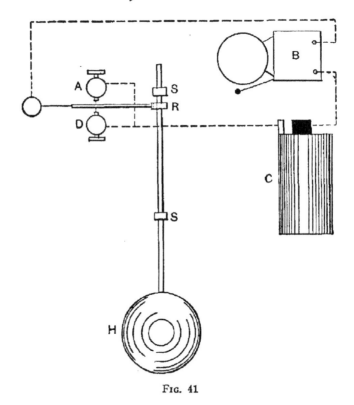

FIG. 41

Another and simpler form.is shown in Fig. 42, where the ball H is mounted on the end of a lever L pivoted at P, its rise or fall completing the circuit of B and C as before.

Where it is desired to give a different signal for
the rise and the fall of level, two bells B and E

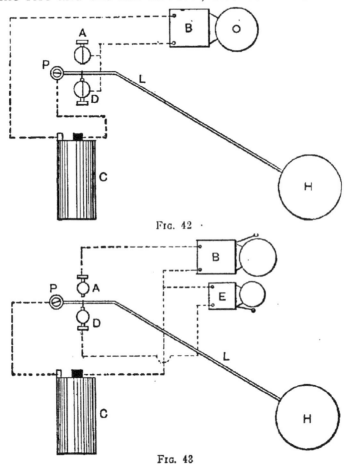

Fig. 42

Fig. 43

(Fig. 43) may be used connected as shown. The
rising of the ball will ring bell B, and its fall,
bell E.

In both forms of indicator, a means must be provided that an undue rise may not bend the lever. This may be accomplished by using contact springs instead of contact screws; it is, however, then harder to adjust the indicator to fine differences of level.

In all cases the contacts must be faced with platinum to prevent corrosion.

Burglar Alarms. A burglar alarm is a device for indicating the opening of a door or window, by the ringing of a bell or operation of an annunciator. The contact apparatus at the points to be protected may either open an electrical circuit or close one, in the latter case being mere modifications of push buttons. The simplest form is the latter or open-circuit method.

The spring contact to be inserted in the door jamb or window frame is so constructed that while under pressure the contacts are kept apart and the circuit is open. But when the door or window is opened, the pressure is released and a spring forces the contacts together.

Fig. 44 is an open-circuit window spring fitted in the window frame so that when the window is closed the spring lug S is pressed inwards, breaking contact with the base B.

If the window is raised, the lug flies to the

position shown by the dotted lines, and making contact with *B*, completes the circuit through bell and battery. These springs are fitted in the side

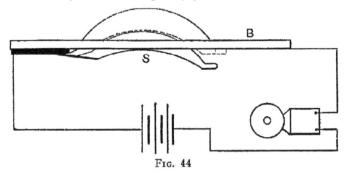

Fig. 44

of the window frame in a vertical position and are entirely concealed when the window is shut.

In the closed-circuit system the reverse happens. The pressure of the closed door or win-

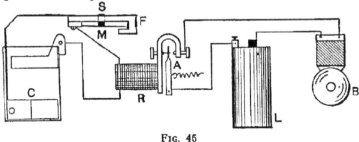

Fig. 45

dow keeps the contacts together and its opening enables them to spring apart.

In Fig. 45 is a diagram of a closed-circuit burglar alarm, *C* a cell of gravity battery, *R* a relay, F the fixed contact and M the movable contact of the spring, S a stud projecting through the base of the spring and pushed in by the closed door.

When the door is closed, S being pushed in,

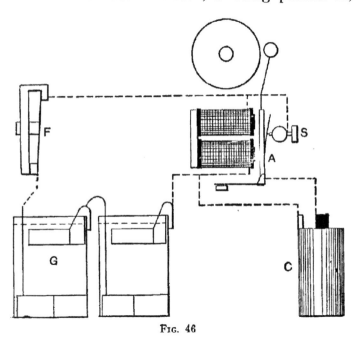

Fig. 46

the circuit of C, R, F and M is closed. The magnets of the relay hold the armature arm A forward against a hard rubber contact. But when S is released, the relay circuit is opened, R loses its power and A flies back, making contact, and throwing in circuit bell B and battery L.

A form of bell and relay combined is shown in
Fig. 46. Here the armature *A* is held against the
magnets while the circuit through the spring *F*
and battery *G* is closed. But on opening this circuit the armature flies back and makes contact
w;th an adjustable contact screw *S* putting in
circuit a local battery *C*. The bell is now practi-

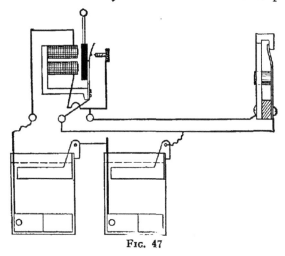

Fig. 47

cally a vibrating bell; on a closed circuit it rings
until the circuit is again closed or the battery runs
down.

A different connection of the same scheme is
Fig. 47, where only one battery is used. This
must be a gravity battery or some other closed-
circuit battery. The circuit can be easily traced
in the figure and needs no special description.

Both of the latter schemes are inferior to one using a separate relay. If the circuit at the spring were quickly closed again the bell would either stop ringing, or be so hampered as to ring very weakly.

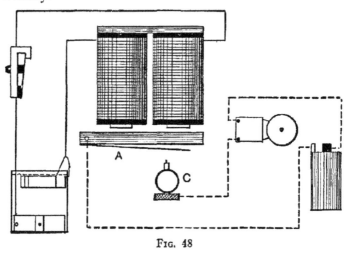

Fig. 48

A relay made as in Fig. 48 has no spring support to the armature A, which falls down by gravity. The adjustable contact C is screwed far back, so that the armature must fall a considerable distance away from the electromagnets before it makes contact. This ensures that the armature will not be attracted and the bell stopped from ringing by a re-closing of the circuit at the door or window spring.

A shade spring (Fig. 49), is made for either

open or closed circuits. In operation, the shade is
pulled down and its string or ring hooked on
to *H*. This draws *H* up a trifle against a spiral
spring and its lower end makes contact with an
insulated spring *S* closing the circuit. If the
shade is disturbed, the spiral spring on the lower
part of *H* is released and it causes a break of
contact with *S* in the direction of the arrow.

When made for open circuit, *S* is bent so that

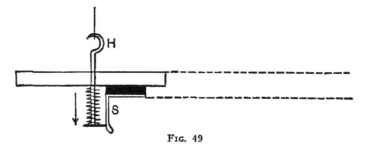

FIG. 49

while under tension no contact is made, but release of tension causes the contact.

Fig. 50 gives the wiring of two windows and
a door on the closed-circuit system. It will be
seen that the contact springs are all in series,
opening a window or the door will thus break
the circuit.

When setting the alarm at night by connecting
up the batteries, relay and bell, should any one
of these springs be open the relay armature will
not hold, and the bell rings.

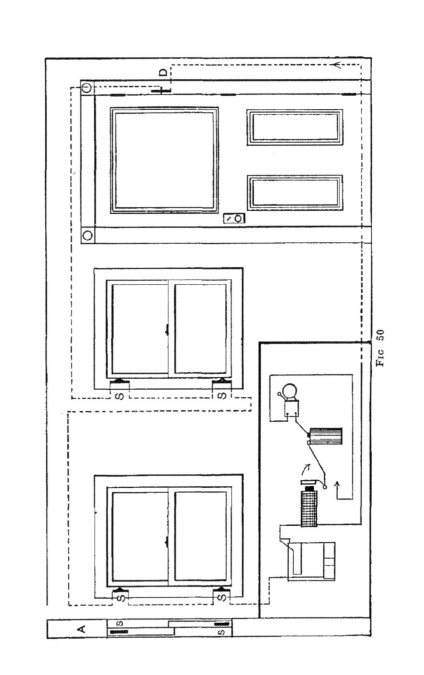

Fig. 50

In this figure the relay is replaced by an electro-magnet holding up a drop shutter by magnetic attraction. Upon the circuit opening, this shutter falls, exposing a number painted on it. At the same time it hits a spring contact placed below it and closes the bell and local battery circuit.

Door Trip Alarm. A swinging contact door trip can be attached over a door to ring a bell when the door is opened.

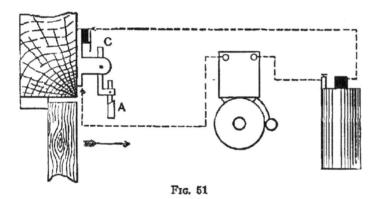

FIG. 51

In Fig. 51 the door trip is screwed over the door so that the lowest arm A is struck by the door. When the door is opened, in the direction of the arrow, the arm A is thrust forwards, and in its turn moves the contact arm C, completing the bell and battery circuit. But when the door is being closed, A swinging in the reverse direction does not move C and no alarm is given.

Miscellaneous Alarms. The Applegate electrical matting is composed of wooden slats with springs so arranged that the weight of any person stepping on it will close a circuit and ring a bell.

It is intended to be put under the ordinary door mat or under stair and room carpeting.

The Yale lock switch is a Yale lock and switch combined. Upon any key but the right one being inserted, a circuit is closed and an alarm bell is rung.

CHAPTER V

Annunciators

The Annunciator. The mechanism of an annunciator consists of electromagnets which allow shutters to drop or needles to move on the circuits being closed. A bell is also rung in most cases to call attention to the annunciator. The number of the circuit is marked on the shutter,

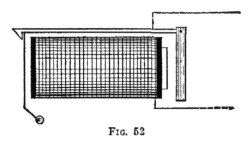

Fig. 52

or near the needle, either shutter or needle being replaced by a reset device, which may be mechanical or electrical.

Annunciator drops are made in a variety of forms. Fig. 52 illustrates the principle underlying nearly all of them.

When current flows through the magnet coils M, the armature A is attracted, and being pivoted at P, the lever hook H rises and allows the weighted shutter S to fall and display a number painted on its inside surface.

The needle drop in Fig. 53 is one that has met with great favor and works as follows: the soft iron core of the magnet C has a hole drilled through it, in which turns the shaft S. An arrow or needle is attached at the front end over the

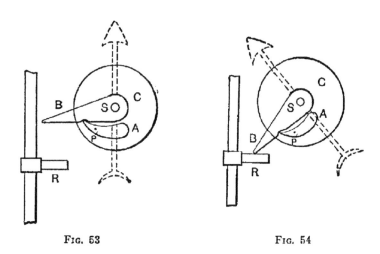

Fig. 53   Fig. 54

face of the annunciator. A notched arm B is fixed on the rear end of the shaft and is held in a horizontal position by the end of armature A.

When the current flows around C, armature A turns on its pivot towards the core of C, as in Fig. 54, unlocking B, which falls and thereby partly rotates shaft S and the arrow.

When it is desired to reset the arrow and arm,

a button is pressed upwards, which raises a rod carrying an arm *R*. This latter arm in turn raises *B* to its former position, the heavy end of *A* falls, and its pointed end locks *B*.

Pendulum, or swinging, signals are used in annunciator work, where there is a liability that the

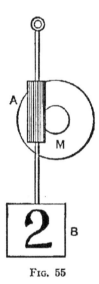

Fig. 55

ordinary drop shutter would not be reset. They, however, only give a visible signal for a few seconds, and are therefore liable to be overlooked.

In Fig. 55 a pivoted arm carrying a soft iron armature *A* and a thin plate *B* having a number on it is free to swing in front of an electromagnet *M*.

When the current flows in the electromagnet the armature is attracted, and upon the circuit being broken at the push, the armature is released and the arm swings to and fro.

The drops of an annunciator are wired up as in Fig. 56.

One end of each coil is attached to a common

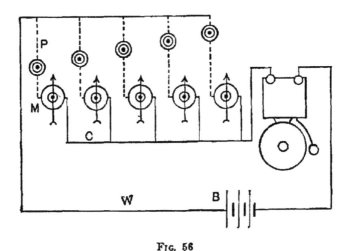

FIG. 56

return wire C, the other end going to the push P. When P is depressed, the circuit of any drop is through M along C through bell, battery and up common battery wire W back to other contact of push P. Depressing any push does not therefore affect any other drop but the one controlled by it.

Wiring up an Annunciator. A diagram of the connections for an annunciator with a separate bell is given in Fig. 57. Where the bell is contained in the case a terminal will be generally found for connection.

The figure shows a wire running from the battery to one side of each push button. This is the common return, or battery wire, and saves instal-

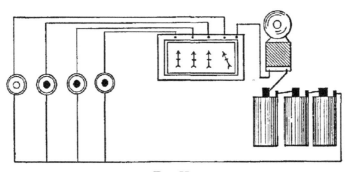

FIG. 57

ling two wires from each push. It should be larger, however, than the rest of the wires, generally about No. 16 B. & S.

All the wires for an annunciator should be run before connecting up. There are different methods of sorting out the wires at the annunciator. One way is to connect the wires (except of course common or battery return wires) to the drops in any order. Then an assistant travels from push to push and presses each button, noting the room numbers and the order in which they were visited.

As each drop falls, its number and order is noted.

Comparing this with the list made by the assistant will show the correct changes to make.

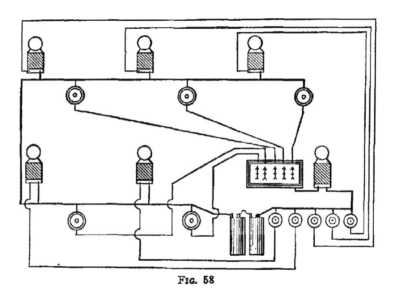

Fig. 58

For instance, suppose pushes 1, 2, 3, 4, 5 and 6 were pressed in that order, and drops 3, 4, 5, 1, 2 and 6 fell in that order. Then the wires at the annunciator would be changed as follows: From 3 to 1, 4 to 2, 5 to 3, 1 to 4, and 2 to 5; 6 would already be in its right place.

Another way is to commence by twisting together say the wires at No. 1 push. Then go to the annunciator and touch each of the push wires to No. 1 drop until it falls. Then connect it, untwist the wires at No. 1, push and connect it up. Proceed to No. 2 and so on until all the pushes have been connected in turn.

In some cases it is desired to answer back to the person calling, or to be able to call any person from the annunciator.

A circuit like Fig. 58 answers the purpose of both annunciator call and return, or fire, call. This requires two wires from each room to the annunciator and a common return wire. By tracing out the circuit it will be seen that when a room push is pressed, the annunciator needle and bell indicate. And when one of the pushes near the annunciator is pressed, the corresponding room bell rings. The former circuit is from the push, along the common return wire, through bell and annunciator back to the push.

The fire call is from push up line to bell through bell along common return and through battery to the push.

The Western Electric single-wire system (Fig. 59) uses three-point pushes, two batteries and two return wires. Battery A is for the annunciator circuit and battery F for the fire, or return, call.

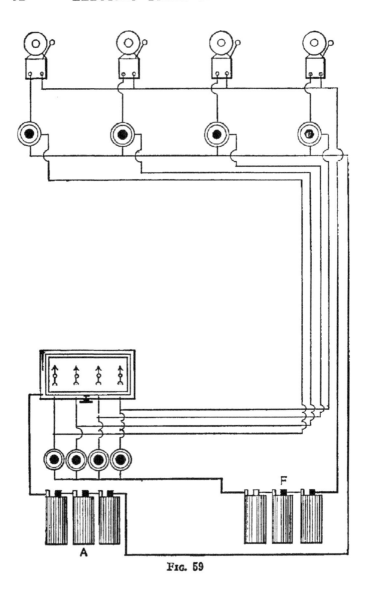

FIG. 59

In each room the top contact and push spring contact are normally together.

If one of the pushes below the annunciator is pressed, battery F is thrown in series with the bell in the room.

But when the room push is pressed its bell is cut out and the circuit becomes like an ordinary annunciator circuit.

CHAPTER VI

Annunciators and Alarms

Three Wire Return Call System. A three wire return call annunciator system is shown in Fig. 60.

There are two battery wires installed, from which taps are taken off and led to each room or push button.

Three way or return call push buttons are used as shown at points marked B.

In the diagram, the bells are marked A, the drops in the annunciator D, the annunciator bell C and the return call buttons in the annunciator E. The batteries are as shown at F. The heavy black outline encloses the annunciator mechanism and connections which are drawn diagrammatically for the sake of clearness.

Three stations only are shown on the sketch, but the annunciators which are manufactured by Edwards and Co , Inc., of New York, are made in all standard sizes.

Installing Elevator Annunciators. The installing of electric bells and annunciators in elevators does not present any special problems, although the apparatus used must be selected with a view to

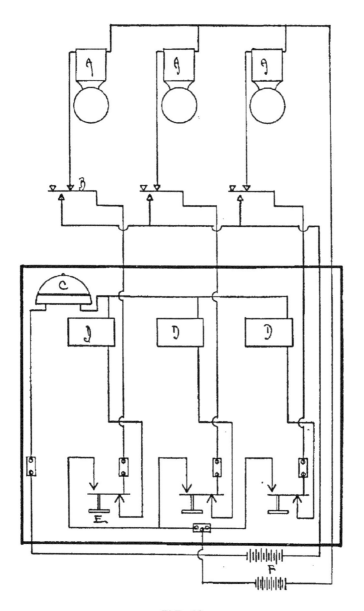

FIG. 60

its being suitable to withstand the shocks incident to elevator service.

In general the wires leading from the push buttons on the different floors to the bell or annunciator in the elevator, are flexible and made up into a cable. One end of this cable is attached to the underside of the elevator car, the other end being fixed usually to the elevator wall, at a point midway between the top and bottom of the shaft.

In Fig. 61 is shown a diagram of the general circuit used, details of course differing in each installation.

One point to be taken care of in elevator work is the attachment of the cables. The continual movement tends to break the wires at the two ends if good flexible cable is not used and the installation done in a workmanlike manner.

Elevator cable is a standard article and may be procured through any electrical supply store. That most commonly used consists of the requisite number of copper conductors each composed of 16 strands' No. 30 B and S. gauge soft and untinned copper wire. These flexible conductors are insulated with two reverse wrappings of cotton and one braid of cotton. The insulated conductors are cabled together with a steel supporting strand where extra tensile strength is required, as in the case of extra long cables. The number of conductors generally ranges from 3 to 20 inclusive.

The wires leading from the push buttons to the cable should be preferably rubber covered and

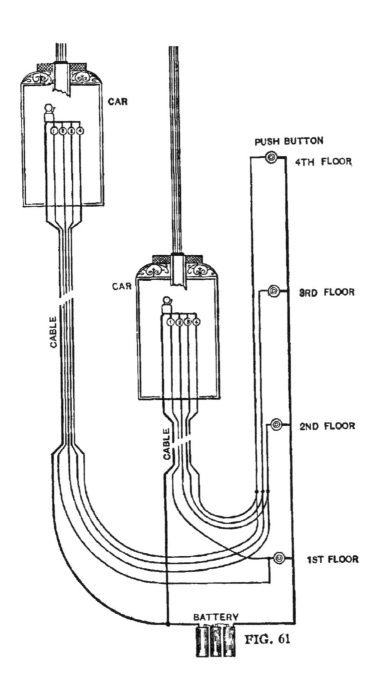

FIG. 61

braided. Only where economy at the outset is desired may ordinary annunciator or office wires be employed.

A connection block carrying binding posts is used at each point where the cable connects to the push button wires or to the annunciator. This may be home-made or purchased ready made, as desired

Burglar Alarm Annunciators. Although almost any annunciator may be used for open circuit burglar alarm work, they usually do not contain certain devices which are desirable in burglar alarm work.

In Fig. 62 is shown a diagram of a burglar alarm annunciator, the view being schematic of the back board.

The references are as follows: A is the main alarm bell situated wherever desired and connected to the binding posts BB. The battery connection leading directly to the battery K is marked C and that leading to the contact spring is marked D. The cut-off switch E cuts off the battery while F is the constant ring switch. G is the upper bar and H the lower bar, while the letters JJ denote the indicating drops. The door and window springs are lettered S. At L is a switch which may be used to disconnect the entire burglar alarm system. Where it is desired to disconnect only a section at a time, the switch corresponding to the section is turned off the upper bar G and on to the lower bar H.

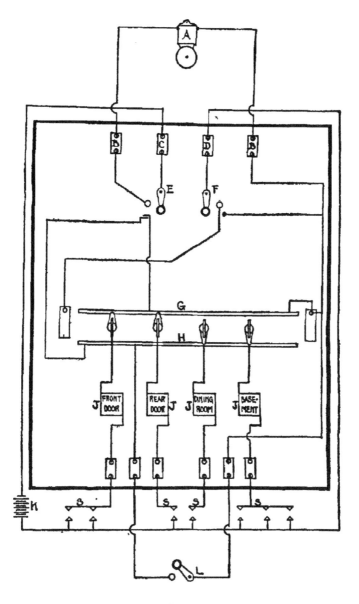

FIG. 62

Clock Alarm Circuit.　A diagram of the wiring and connections on the back board of all clock alarms is illustrated in Fig. 63　This diagram embodies the principles of the last described circuit, but includes the circuit of a clock-operated alarm.

Bells for High Voltages.　The use of electric bells on lighting circuits is becoming quite general, as it obviates the necessity of using batteries, and thereby simplifies both installation and maintenance

There is no fundamental objection to operating make and break bells on electric light circuits.　Providing the voltage and amperage are the same, there is little difference between the current from a direct-current dynamo and that from a battery But owing to the higher voltages of the lighting circuit over that generally employed from batteries. the bell coils must be wound to high resistances to keep down the current strength.　There are also other slight changes to assist in suppressing sparking, as have been already treated on.

Where the circuit is not over 220 volts, the bells are wound with fine wire and have also self-contained resistance coils.　For 500 volts and over, a resistance lamp is connected in with the bell which in this case is wound for a 150-volt circuit.

These bells up to 6-inch and inclusive will operate on circuits of either direct or alternating current.

Above this size it is necessary to use specially constructed bells on alternating current circuits.

Most large hotels and office buildings having

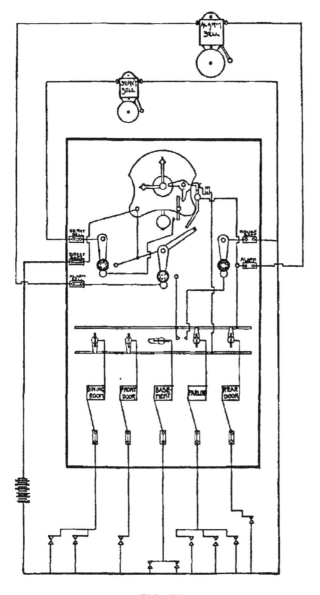

FIG. 63

direct current lighting service are using it for ringing bells and similar work to the total exclusion of batteries.

Where the number of units to be operated justifies it, motor generators are operated in connection with the lighting mains to produce a low voltage most suitable for the bells. The connections in this case are no different to those when batteries are employed.

Bell-ringing Transformers. The best system for operating bells and annunciators from alternating current circuits is undoubtedly that employing small specially constructed transformers to reduce the voltage. These transformers are being used universally for hotel and office work where alternating current is available. They are simple, being merely one or more coils of well insulated wire wound on soft iron cores and having connections for both the lighting circuit and the bell circuit.

As a general rule the coils are divided as to their number of turns or according to the ratio of transformation desired. For example, if the circuit were 110 volts and 10 volts was required for the bell circuit, the total number of turns in the transformer would be connected, $\frac{10}{11}$ to the lighting circuit and $\frac{1}{11}$ to the bell circuit.

The bell-ringing transformers on the market are made in several styles. One small style, Fig. 64, for single residences, is for use on 110 volts and produces a bell voltage or secondary voltage as it

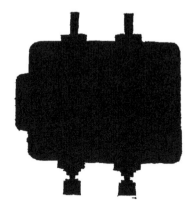

FIG. 64

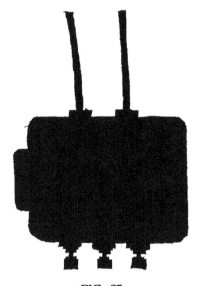

:FIG. 65

BELL-RINGING TRANSFORMERS.

is termed, of 6 volts. Another size, Fig. 65, of this transformer has three secondary voltages 6, 12 and 18, each of which can be used by connecting to the right binding posts.

It is to be noted that where the lighting service voltage or primary voltage varies from the above, the secondary voltage delivered to the bell circuit will vary in like proportion. It should also be noted that a careless reversing of the connections, that is connecting the secondary leads to the lighting circuits, instead of the primary leads would cause a like high voltage at the other terminals of the transformer, raising it in due proportion instead of lowering it. Thus such carelessness would produce a voltage of 2,400 volts instead of 6 if a transformer intended to deliver 6 volts from a 120-volt circuit was wrongly connected.

The results might very well then be dangerous. All transformers are properly marked, however, and such an error only occurs through ignorance or carelessness.

The installation of these bell-ringing transformers is simplicity itself; they require no care after installation and have met with the approval of the National Board of Fire Underwriters.

Combination Circuits. Circuits intended primarily for electric bells or annunciators in houses and apartments may often be also made to serve for other electrical devices such as door openers, house telephones, etc. This subsidiary apparatus

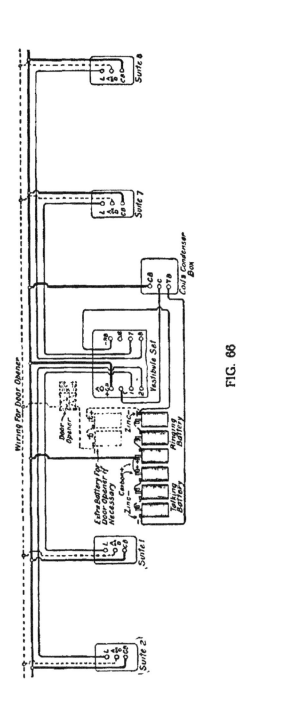

FIG. 66

may be installed with a little additional wiring or perhaps will not need any other wires, as when both the devices are not used at once.

Electrical door openers are great conveniences and are practically indispensable where the outside door is on another level to the location of the dweller or where two or more families occupy the same house. The device is simple, consisting of an electrically released spring-plate against which the lock bolt is normally held and a door opening spring

When the door opener button is pressed, the spring plate is released, releasing the lock bolt by the same action. The door spring then forces the door open enough to clear the opener plate, which flies back into position when the button is released.

These door openers are made in several forms for door frames, such as those on thin doors, iron gates, for surface or rim locks, for thick doors, sliding doors and any other regular type of door.

The push button is the same as used for electric bells and may be located wherever desired. The pushes are wired in multiple as shown in Figs. 66 and 67, which are two circuits of a type of the Western Electric interphone, a system of house telephones supplied for houses and buildings of every size. Fig. 66 shows a circuit which provides telephone service between the vestibule and the apartments, the door opener wiring being clearly indicated. In Fig. 67 the circuit provides a more extensive service, enabling the janitor, the apartments and the tradesmen to intercommunicate in

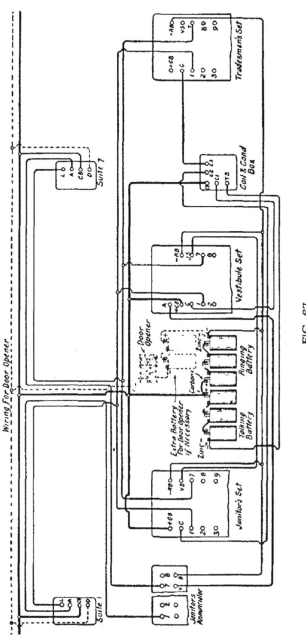

Wiring For Door Opener

Suite 7

Suite 1

Tradesmen's Set

Coil & Cond. Box

Vestibule Set

Extra Battery
For Door Opener
if necessary

For your Door
Opener

Ringing
Battery

Talking
Battery

Janitor's Set

Janitor's
Annunciator

FIG. 67

the most desirable system. The door opener wiring is also clearly shown.

The convenience of having telephone connection in the house or hotel and its advantages over speaking tubes are too well known to need extended comment. Where electric bells have already been installed it is quite feasible now to use the same wires for telephones also.

Telephone sets especially designed for this service are manufactured by the Western Electric Company in their interphone series. They are simple and compact, and may be installed by anyone who can put up an electric bell.

Fire Alarm Circuits. A fire alarm circuit suitable for factories, private plants or groups of buildings is shown in Fig. 68. It is a series system, with closed circuit, the gongs sounding whenever the circuit is opened whether by the contact breaker in the boxes or by the accidental breaking of a wire. This insures that it remains in good working order, as when any part of the circuit is opened, a warning tap is sounded on every bell or gong.

The boxes have contact breakers which send a separate number of impulses for each box, thus announcing the box number on each gong. The boxes and gongs may be located anywhere, as the system is perfectly flexible.

The reference letters in the diagram are as follows: *C* indicates the gongs which are preferably of the electro-mechanical type, a coiled spring pro-

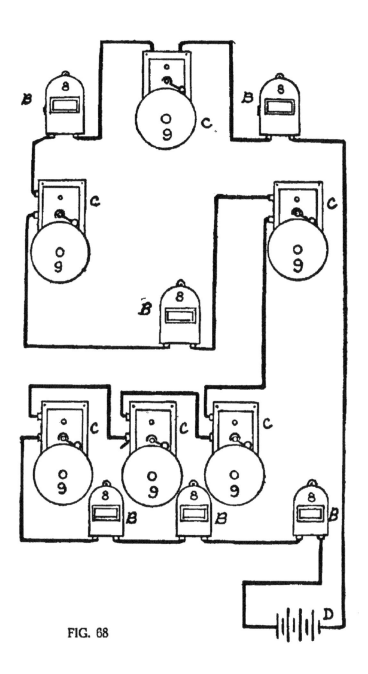

FIG. 68

viding force for the blow, electricity being merely used to release and retain the hammer or striker. The alarm boxes are marked *BB* and the battery which is of the closed circuit type is marked *D*.

Interior Fire Alarm System. Another system suitable more particularly for indoor operation is illustrated in Fig. 69. Here the alarm is given by breaking the glass front of an alarm box and releasing or pressing an electrical contact.

The box sounded indicates by causing a drop to fall on an annunciator and at the same time rings an alarm bell. The latter are generally provided with constant ring attachments, which keep the bell sounding until shut off.

The annunciator shown in the diagram has switches for controlling each individual bell circuit, and also for control of the entire system.

There is no practical limit to the number of stations in this system, it being determined by the size of the annunciator used or by other obvious factors.

The reference letters on the diagram are as follows: *A*, alarm bells which may be located wherever desired. *B*, break-glass alarm boxes also located at convenient points. *C*, annunciator drops, *D*, switches on annunciator which control each individual bell circuit, enabling any circuit to be cut out, cut in or tested without disturbing any other circuit. *E* is a general alarm switch, causing all bells to ring at once when it is operated.

The battery *F* varies with the number of bells and

boxes and the length of line, from three cells upwards   A cut-out switch *H* is added to cut out
the entire system by opening the battery wire.

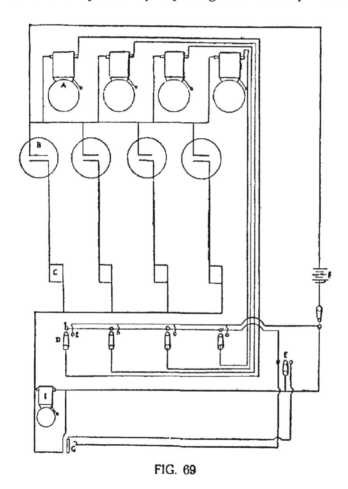

FIG. 69

The annunciator bell is at *I*, an auxiliary bell being
added in multiple with it when necessary.

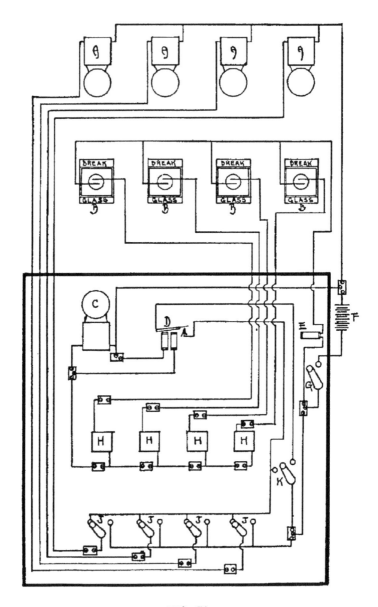

FIG. 70

Fire Alarm System for Considerable Areas.
Where the area is more extensive and the number
of stations considerable, the system illustrated in
Fig. 70 is very suitable. It consists of the requisite
number of break-glass boxes, bells and a more
elaborate annunciator system. In general details it
resembles the last system, but uses a relay to send
out the current for ringing the alarm bells.

When a box operates, the current impulses sent
on the line act on the relay instead of directly on
the bells. Each stroke of the relay closes a local
circuit which includes the bells and the battery.

This system does away with large batteries and
is very enconomical of wire . The current needed
for the relay is very small, whereas in a direct system of any size, the current and voltage to ring a
number of bells located at wide intervals would be
prohibitive.

The reference letters are as follows: AA are
the alarm bells, BB the break-glass alarm boxes,
C is the annunciator bell, D is the relay which remains closed when an alarm comes in keeping the
bells constantly ringing until shut off. E is a resistance coil and F is the battery.

A system cut-out switch G and JJ switches on
the annunciator for controlling individual circuits
are also provided. HH are the annunciator drops
and K is a constant-ring switch which can also be
used for a general alarm to ring all the bells at
once.

CPSIA information can be obtained
at www.ICGtesting.com
Printed in the USA
LVOW04*2127131115

462477LV00013B/258/P